Mycology
for CLS and MLT

Order this book online at www.trafford.com
or email orders@trafford.com

Most Trafford titles are also available at major online book retailers.

Printed in the United States of America.

ISBN: 978-1-4269-7904-0

Trafford rev. 07/26/11

www.trafford.com

North America & international
toll-free: 1 888 232 4444 (USA & Canada)
phone: 250 383 6864 ♦ fax: 812 355 4082

Mycology

For CLS & MLT

Mary Michelle Shodja, MS, MT (ASCP), CLS

Contents

List of Figures   v

List of Pictures   v

List of Tables   vii

Preface   ix

SECTION I: Introduction to Mycology   1

SECTION II: Sources And Common Opportunistic Pathogens Isolated   8

SECTION III: Mycology Media   10

SECTION IV: Specimen Processing   11

SECTION V: Fungal Identification   13

SECTION VI: Medically Important Fungi   50

SECTION VII: Safety   51

SECTION VIII: Antifungal Susceptibility Test (AST)   52

MYCOLOGY -WRITTEN EXAMINATION   54

Mycology Written Exam Answer Key   63

Afterword   64

Glossary   65

References   67

About the Author   68

List of Figures

Figure 1 – Medically Important Phyla of Fungi   3

List of Pictures

Picture 1- Zygospores   4
Picture 2 – Ascospores   5
Picture 3 - Basidiospores   6
Picture 4 - Yeast on Saline Wet Mount   13
Picture 5 - Yeast culture (pasty-look)   14
Picture 6 – C. neoformans, india ink + (50X)   14
Picture 7 - Germ Tube positive Candida albicans (50X)   15
Picture 8 - C. albicans on cornmeal agar (50X)   16
Picture 9 - C. glabrata on cornmeal agar (50X)   16
Picture 10 - C. tropicalis on cornmeal agar (50X)   16
Picture 11 - C. krusei on cornmeal agar (50X)   16
Picture 12 - C. parapsilopsis on cornmeal agar (50X)   16
Picture 13 - Saksinaea vasiformis culture   18
Picture 14 – A. fumigatus (50X)   19
Picture 15 -A. niger (50X)   19
Picture 16 - A. flavus (50X)   19
Picture 17 - A. Terreus   19
Picture 18 - Aspergillus colony morphologies   19
Picture 19 - Fusarium oxysporum on PDA   20
Picture 20 - Fusarium verticillioides (50X)   20
Picture 21 - Alternaria spp. culture   21
Picture 22 - Alternaria spp. culture   21
Picture 23 - Alternaria spp. (50X)   21
Picture 24 - Alternaria spp. (50X)   21
Picture 25 – Malassezia furfur culture   22
Picture 26 – Malassezia furfur 50X   22

Picture 27 – Ptyriasis versicolor   22

Picture 28 – Exophiala werneckii culture   23

Picture 29 – Exophiala werneckii 50X   23

Picture 30 – Tinea nigra palmaris   23

Picture 31 – Piedraia hortae/nigra culture   24

Picture 32 – Piedraia hortae microscopic   24

Picture 33 – Hair nodule infected with P. hortae   24

Picture 34 – Trichosporon spp. culture   25

Picture 35 – Trichosporon beigelii 50X   25

Picture 36 – White piedra of the hair nodule   25

Picture 37 – Microsporum canis 50X   26

Picture 38 – Microsporum gypseum 50X   26

Picture 39 – Microscporum audouinii 50X   26

Picture 40 – Microsporum canis culture   27

Picture 41 – Epidermophyton flocossum 50X   27

Picture 42 – Epidermophyton flocossum culture   28

Picture 43 – Trichophyton mentagrophytes culture   28

Picture 44 – T. mentagrophytes (50X)   29

Picture 45 – Trichophyton rubrum (50X)   29

Picture 46 – Trichophyton mentagrophytes (50X)   30

Picture 47 – Trichophyton violaceum (50X)   30

Picture 48 – Trichophyton schoenleinii (50X)   31

Picture 49 – Trichophyton verrucosum (50X)   31

Picture 50 – Sporothrix schenckii culture   32

Picture 51 – Sporothrix schenckii microscopic mold form (50X)   32

Picture 52 – Sporothrix schenckii microscopic yeast form (50X)   33

Picture 53 – Acremonium spp. culture   33

Picture 54 – Acremonium spp. (50X)   34

Picture 55 – Exophiala jeanselmei culture   34

Picture 56 – Exophiala jeanselmei (50X)   35

Picture 57 – Cladosporium carrionii culture   36

Picture 58 – Cladosporium spp. (50X)   36

Picture 59 – Phialophora spp. culture   37

Picture 60 – Phialophora spp. (50X)   37

Picture 61 – Fonsecaea pedrosoi culture   38

Picture 62 – Fonsecaea pedrosoi (10X)   39

Picture 63 – Fonsecaea pedrosoi conidia (50X)   39

Picture 64 – Fonsecaea pedrosoi conidia (50X)   40

Picture 65 – Alternaria spp. culture   40

Picture 66 – Alternaria spp. (50X)   41

Picture 67 – Wangiella dermatitidis culture   41

Picture 68 – Wangiella dermatitidis (50X)   41

Picture 69 – Coccidiodes immits mold culture   42

Picture 70 – Coccidiodes immitis mold form (50x)   42

Picture 71 – Coccidiodes immitis yeast in tissue (100x)   43

Picture 72 – Histoplasma capsulatum culture   44

Picture 73 – Histoplasma capsulatum mold (50X)   44

Picture 74 – Histoplasma capsulatum yeast (100x)   44

Picture 75 – Blastomyces dermatitidis culture   45

Picture 76 - Blastomyces dermatitidis mold (50X)   46

Picture 77 - Blastomyces dermatitidis yeast (100X)   46

Picture 78 – Paracoccidiodes brasiliensis culture   47

Picture 79 – Paracoccidiodes brasilienses mold (50X)   47

Picture 80 – Paracoccidiodes brasiliensis yeast (100X)   48

Picture 81 - GMS positive Pneumocystis jiroveci   49

List of Tables

Table 1 – Sources and Common Opportunistic Pathogens isolated   10

To my mother, for her sacrifices, this is for you.

To my youngest sister Luanne Freeman, I am so proud of the person you had become, I was lucky to have a small part in raising you because now I can nicely pat myself on the back for a job well done. With deep gratitude to my cousins Chona Aros and Christine Sy. Lastly, to my classmate and best friend, Anna Hamilton, thank you for your friendship of over 20 years.

Author picture is courtesy of Marjorie Reyno, MS, MT, (ASCP-M), CLS, Microbiology Manager, Loma Linda University Medical Center, Loma Linda, California.

"In the end, it's not the years in your life that count.
It's the life in your years."
Abraham Lincoln

To laboratory students and fellow laboratorians, never forget what you learned and never stop learning new things

The Author

Preface

This manual is the fourth of a series of 20 manuals that the author was commissioned to write for the Medical Laboratory Technician (MLT) training program in Diamond Bar, California that was granted approval as a training facility in 2009 by the State Department of Health and Human Services. In writing these manuals, the author strived to adhere to the strict guidelines of the State of California's requirements for the MLT program. At the time of publication, the initial beneficiary of these manuals, Diamond Bar California's first MLT graduate, successfully passed the ASCP examination.

The author revised these manuals to serve 3 purposes: as the primary textbooks for the 6-month MLT training program, as reviewers for Clinical Laboratory Scientists (CLSs) preparing for the CLS Licensure or Certification, and as continuing education materials for CLSs and MLTs as a requirement for license renewal.

The author wrote these manuals in an outline form for easy reading and understanding and free from the constraint of a formal textbook. The author's intention is to speak to the reader from the actual clinical laboratory bench than from the classroom.

The Mycology for CLS and MLT only discussed the most common fungus isolated in the clinical laboratory and is not recommended as a replacement of the actual textbooks currently being used in the CLS program.

Also by Mary Michelle Shodja, MS, MT (ASCP), CLS:

Bacteriology for CLS & MLT
Hematology for CLS & MLT
Parasitology for CLS & MLT
Virology for CLS & MLT
Coagulation for CLS & MLT
Urinalysis & Body Fluids for CLS & MLT
Routine Chemistry for CLS & MLT
Special Chemistry for CLS & MLT
Toxicology for CLS & MLT
Serology - Immunoassays for CLS & MLT
Serology - ELISA Assays for CLS & MLT
Serology & Syphilis for CLS & MLT
Immunology for CLS & MLT
Blood Bank for CLS & MLT
Phlebotomy for CLS, MLT & Phlebotomist
Specimen Processing for CLS, MLT & Phlebotomist
Laboratory Safety for CLS, MLT & Phlebotomist
Total Quality Management I – Quality Assurance
Total Quality Management II – Quality Control

SECTION I:

Introduction to Mycology

FUNGI

- Or Kingdom Fungi
- A Eukaryote
- Has approximately 40,000 different kinds including yeasts and molds
- Cell wall contains a substance called chitin
- Causes a number of nosocomial infections
- Chemoheterotrophs – require organic compounds for energy and carbon
- Few are anaerobes
- Major decomposer of dead plant matter through enzyme cellulase
- Humans use for food (mushroom, production of bread and citric acid) and drugs (alcohol and penicillin)
- Mycology – study of fungi

CHARACTERISTICS OF FUNGI
- Vegetative structures – fungal colonies are composed of cells involved in catabolism and growth
- Molds and Fleshy Fungi
  - Thallus (body) – consists of long filaments of cells joined together
  - Hyphae – long filaments, can grow to immense proportions
  - Septa – cross-walls which divide the hyphae into distinct uninucleate cell-like units, "septate hyphae"
  - Coenocytic hyphae – no septa
  - Vegetative hypha – portion of hypha that obtains nutrients
  - Reproductive or aerial hypha – portion concerned with reproduction, projects above the surface of the medium on which it is growing, often bearing reproductive spores
  - Mycelium – a filamentous mass which is visible to the unaided eye

- Yeasts – nonfilamentous, unicellular fungi that are typically spherical or oval
  - Widely distributed in nature
  - Frequently found as a white powdery coating on fruits and leaves
  - Capable of facultative anaerobic growth
  - Can use oxygen or an organic compound as the final electron acceptor

- Allows fungi to survive in various environments
- Ferments carbohydrates and produces ethanol and CO_2
- Budding – parent cell forms a protruberance (bud) on its outer surface, 1 yeast cell can produce up to 24 daughter cells by budding
- Pseudohypha – buds that fail to detach forming a chain of cells, Candida albicans attaches to human cells as a yeast but requires pseudohypha to invade deeper tissues

- Dimorphic Fungi – pathogenic species
  - Exhibits dimorphism
  - 2 forms of growth, mold at RT and yeast at 37^0C

FUNGI NUTRITIONAL ADAPTATIONS
- Generally adapted to environments that will be hostile to bacteria
- Chemoheterotrophs, they absorb nutrients rather than ingesting them as animals do
- Grows better at pH of about 5.0
- Almost all are aerobic, most yeasts are facultative anaerobe
- More resistant to osmotic pressure
- Can grow in high sugar and salt
- Can grow on substances with very low moisture content
- Require less nitrogen for equivalent amount of growth
- Capable of metabolizing complex carbohydrates, such as lignin (component of wood)

Figure 1

Medically Important Phyla of Fungi

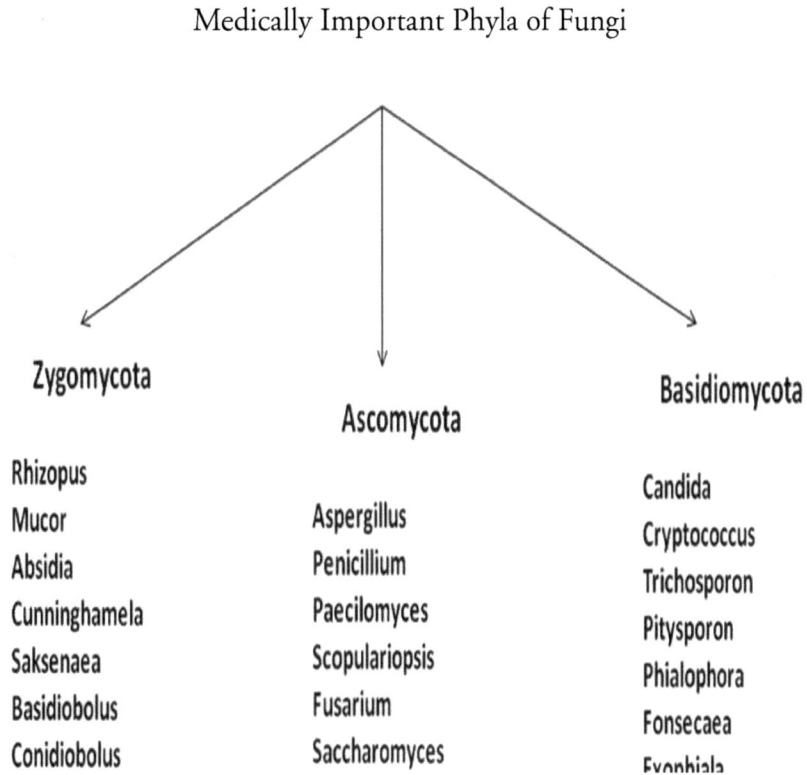

PHYLUM ZYGOMYCOTA
Zygomycota – conjugation fungi, saprophytic molds that have coenocytic hyphae

Zygospores – sexual spores, large, enclosed in a thick wall

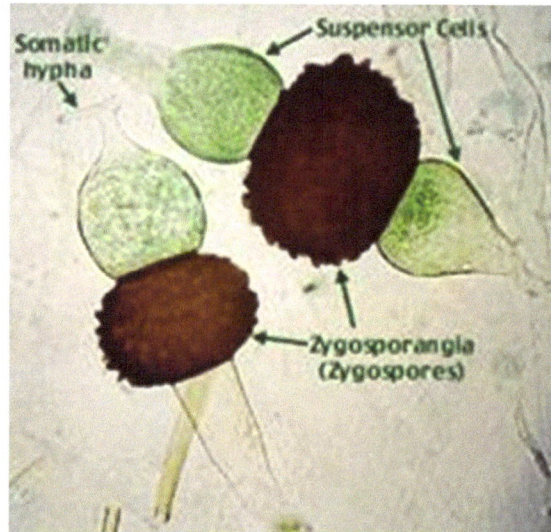

Picture 1- Zygospores

GENERA
- Rhizopus
- Mucor
- Rhizomucor
- Absidia
- Cunninghamela
- Saksenaea
- Basidiobolus
- Conidiobolus

PHYLUM ASCOMYCOTA
Ascomycota or sac fungi – include molds with septate hyphae and some yeasts, asexual spores are usually conidia produced in long chains from the conidiophore, conidia detaches easily from the chain and floats in the air like dust

Ascospore – produced in a sac-like structure called ascus

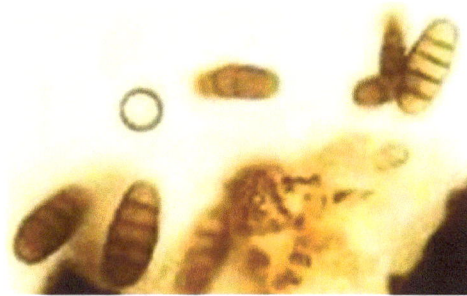

Picture 2 – Ascospores

GENERA

- Aspergillus
- Penicillium
- Paecilomyces
- Scopulariopsis
- Fusarium
- Saccharomyces
- Pichia
- Arthroderma (Trichophyton, Microsporum,
- Epidermophyton)
- Ajellomyces (Histoplasma, Blastomyces)

PHYLUM BASIDIOMYCOTA

Basidiomycota or club fungi – also possesses septate hyphae, includes fungi that produce mushrooms

Basidiospores – formed externally on a base pedestal called a basidium, usually 4 basidiospores per basidium, some basidiomycota produces asexual conidiospores

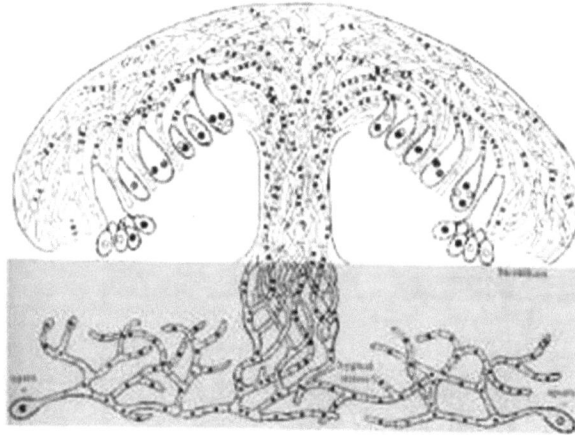

Picture 3 - Basidiospores

GENERA

- Candida
- Cryptococcus
- Trichosporon
- Pityrosporum
- Phialophora
- Fonsecaea
- Exophiala
- Wangiella
- Xylohypha
- Bipolaris
- Alternaria

FUNGAL DISEASES

- Mycosis – fungal infection, generally long-lasting infections because fungi grow slowly, classified into 5 groups according to the degree of tissue involvement and mode of entry into the host (cutaneous, superficial or opportunistic):
- Systemic mycoses – fungal infections deep within the body, usually caused by fungi that live in the soil, inhalation of the spore is the usual mode of transmission
- Subcutaneous mycoses – fungal infections beneath the skin caused by saprophytic fungi that live in soil and on vegetation, usual route is direct implantation of spores or mycelia fragments into a puncture wound on the skin

- Cutaneous mycoses or dermatomycoses – infects only the hair, skin and nails, caused by fungi dermatophytes that secrete keratinase, an enzyme that degrades keratin, a protein found in hair, skin and nails. Transmission is human-to-human or animal-to-human
- Superficial mycoses - fungi are localized along hair shafts and in superficial (surface) epidermal cells, infections are prevalent in tropical climates
- Opportunistic pathogen – generally harmless in normal habitat but can become pathogenic in a host who is seriously debilitated or traumatized, examples are those who are under treatment with broad spectrum antibiotics or immune system suppression by drugs or by an immune disorder or those who have a lung disease

SECTION II: Sources and Common Opportunistic Pathogens Isolated

Table 1 Sources and Common Opportunistic Pathogens isolated

Source	Common Opportunistic Pathogens isolated
Lungs	Cryptococcus neoformans, Scopulariopsis, Coccidiodes immitis, Histoplasma capsulatum, Blastomyces dermatitidis, Paracoccidiodes brasiliensis
Sinus or pulmonary disease	Zygomycosis, Aspergillosis, Alternaria (sinusitis), Coccidioides immitis, Histoplasma capsulatum, Blastomyces dermatitidis, Paracoccidiodes brasiliensis
Gastrointestinal disease	Candida albicans, Zygomycosis
Blood	Candida albicans, Zygomycosis, Coccidioides immitis, Histoplasma capsulatum, Blastomyces dermatitidis
Central Nervous System	Aspergillosis, Coccidiodes immitis, Histoplasma capsulatum, Blastomyces dermatitidis, Paracoccidiodes brasiliensis
Cutaneous	Candida albicans, Aspergillosis, Paecilomyces, Coccidiodes immitis
Hair	Dermatophytes

Continuation Table 1 Sources and Common Opportunistic Pathogens isolated

Source	Common Opportunistic Pathogens isolated
Skin	Dermatophytes
Nails	Scopulariopsis, Dermatophytes
Subcutaneous	Sporothrix schenckii, Acremonium, Exophiala spp., Pseudoallescheria spp., Cladosporium spp., Fonsecaea spp., Phialophora spp., Alternaria spp., Wangiella dermatitidis, Paracoccidiodes brasiliensis (ulcers of mucous membranes)
Endocardial	Aspergillosis, Paecilomyces
Eyes	Paecilomyces, Fusarium
Tissue	Histolasma capsulatum
Bone	Blastomyces dermatitidis
Urogenital	Candida albicans, Blastomyces dermatitidis

SECTION III: Mycology Media

1. Sabouraud's Dextrose Agar (SDA) – acid pH of 5.6 inhibits many bacteria
2. Brain Heart Infusion (BHI) – contains X and V factors, calf brain and other nutrients
3. BHI with 10% sheep blood
4. Media with cyclohexamide (antifungal) and chloramphenicol (inhibits bacteria) – isolates fungi that are resistant to antifungal cyclohexamide – e.g. BHI with C & C
5. Media with C & C and Gentamicin – for those fungi that are resistant to Cyclohexamide and those bacteria that are resistant to Chloramphenicol and Gentamicin
6. Potato Dextrose Agar – made for potato infusion and dextrose (corn sugar) – this media is the most widely used medium for growing fungi
7. Cornmeal Agar – made up of cornmeal and agar – a general purpose medium to cultivate fungi
8. Chromagar® - chromogenic media for the identification of Candida – contains agar, peptone, chromogenic mix and chloramphenicol

SECTION IV: Specimen Processing

TYPES OF SPECIMEN
- Respiratory secretions
1. Sputum
2. Transtracheal aspirates
3. Bronchoscopy
4. Unacceptable specimens:
    - specimens >24 hours
    - swabs of respiratory secretions
    - frozen specimens or specimens transported on ice
- Cerebrospinal Fluid – do not refrigerate, best to concentrate first before inoculating to media
- Blood
- Hair, Skin and Nails
    - Skin –best to get the outer edge of lesion
    - Hair – inoculate directly
    - Nails – grind or place in a liquid media to soften (Thioglycollate Broth or Trypticase Soy Broth)
- Genitourinary tract
- Urine (catheterized if possible), first morning sample preferred
- Swabs from cervix and vaginal
  Unacceptable specimens:
      24-hour urine
      Preserved Urine

- Tissue, bone marrow, and sterile body fluids
    - Bone Marrow – submit in heparinized tube
    - Tissue – specimen should be in sterile saline
    - Sterile body fluids – best to concentrate first inoculating to media
- Wound and Drainage
    - Wound – swab
    - Drainage – specimen in sterile container

SPECIMEN INOCULATION
Primary inoculation media used for fungal growth varies slightly with each laboratory.

The following specimens can be directly inoculated on any of the fungal media listed in Section III, Mycology Media:

1.   Abscess aspirates

2.   Bone marrow aspirates

3.   Biopsy tissues

4.   Hair, skin scrapings and nails

5.   Respiratory specimens

6.   Sterile body fluids

The following specimens need concentration by centrifugation (between 1500-2000 x g for 10 minutes)

1.   Respiratory specimens (if >2.0 ml)

2.   Sterile body fluids (if it is clear)

3.   Urine (if it is clear)

All inoculated fungal media should be sealed with a plate sealer before incubation.

Incubation temperature: 30^0C ($\pm$1)

Length of incubation: 4 weeks – most cultures
6 - weeks for patients suspected of systemic fungal infections, bone marrow and sterile body fluids and tissues

Culture should be inspected for growth at least daily

SECTION IV: Fungal Identification

- Yeast grows faster than molds
- Yeast colonies are opaque and pasty and molds have a "fuzzy" appearance
- Yeast normally grows in simple media like SDA and molds need more complex media with antibiotics because they are slow growers and bacteria will exhaust the nutrients and would inhibit their growth

Yeast

Yeast – unicellular, produce daughter cell from the parent cell by budding. Colonies produce opaque, creamy or pasty colonies

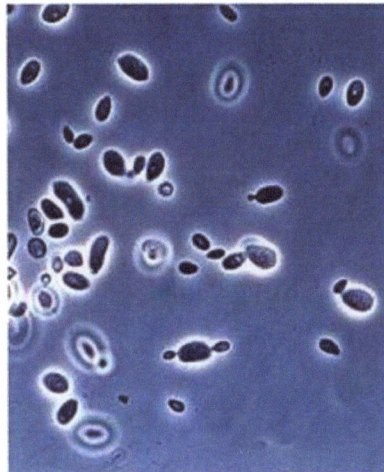

Picture 4 - Yeast on Saline Wet Mount

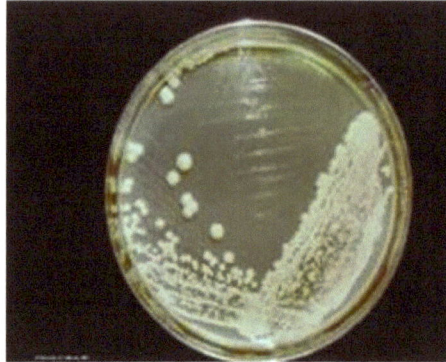

Picture 5 - Yeast culture (pasty-look)

Cryptococcosis and Cryptococcus neoformans
- Found in soil and bird droppings
- 4th most common opportunistic infection among HIV/AIDS and immunocompromised patients
- Laboratory tests:
  - India ink – positive for encapsulated yeast

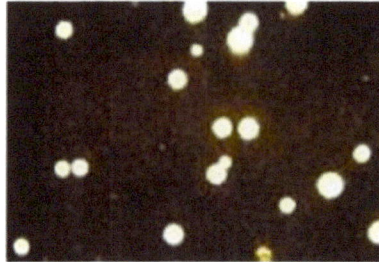

Picture 6 – C. neoformans, india ink +

  - Positive urea, Nitrate negative
  - Niger seed agar – produce maroon-red to brown-black pigment on the agar
  - Yeast ID system

Candida species
- Direct examination
- Presence or absence of pseudohyphae

- Presence or absence of true hyphae and arthroconidia
- Number of buds and nature of attachment to the mother cells
- The 5 most common clinical yeast isolated
  1. Candida albicans
  2. Candida glabrata
  3. Candida tropicalis
  4. Candida krusei
  5. Candida parapsilosis

- Laboratory Tests:
  - Germ Tube Test – culture is added to a suspension of rabbit plasma, incubated for 2-4 hours at 37^{0}C and a wet mount suspension is examined under the microscope. Candida albicans and a rarer Candida dubliniensis will produce a short germ tube.
  -

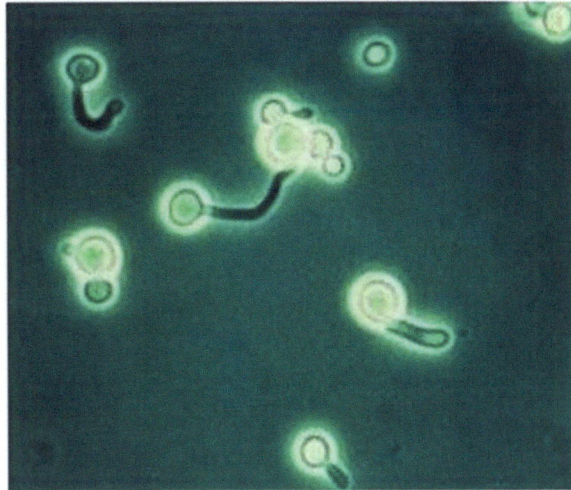

Picture 7 - Germ Tube positive Candida albicans

- Chromagar – various species can be distinguished based on color change on the agar
  - Candida albicans – green
  - Candida parapsilopsis – white-light pinkish
  - Candida glabrata – light pink
  - Candida tropicalis – light purple-metallic blue
  - Candida krusei – dry flat pink
  - Cryptococcus neoformans – white

- Saccharomyces spp. – purple
- Trichosporon spp. – flat light blue

- Cornmeal Agar with or without Tween 80 – enrichment medium which stimulates sporulation of Candida albicans and is useful in suppressing certain other fungal growth

The 5 most common Candida on Corn meal Agar

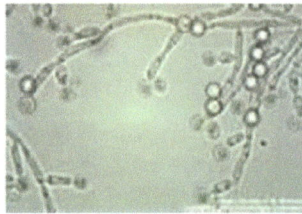

Picture 8 - C. albicans
blastoconidia & chlamydospores)

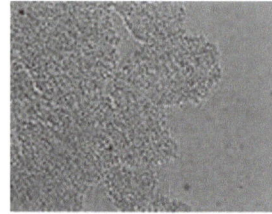

Picture 9 - C. glabrata
((Blastoconidia only)

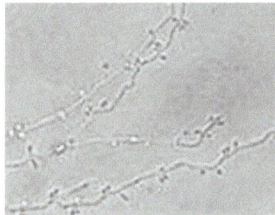

Picture 10 - C. tropicalis
(Blastoconidia sparsely)

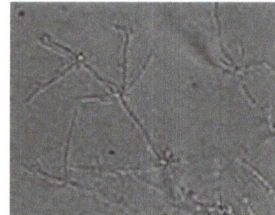

Picture 11 - C. krusei
(Blastoconidia arise only from septa of hyphae)

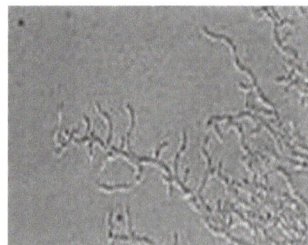

Picture 12 - C. parapsilopsis
(highly branched curved pseudohyphae)

Rapid Urea Test – Cryptococcus neoformans positive

Yeast Identification System

Mold

Methods:
Tape Mount
Tease Mount
Slide Culture
Nutrient Study
Incubation in different temperatures

Mold Identification is based on:
- Specimen source
- Colony morphology
- Growth on BHI with C&C with or without gentamicin
- Morphology on wet mount (lactophenol cotton blue and/or KOH)

ZYGOMYCETES
LABORATORY IDENTIFICATION

COLONY MORPHOLOGY
- Broad, aseptate (occasionally septated) hyphae
- Grows rapidly over the entire surface of the agar plate
- Colony color is gray-white, brown or gray-brown
- Colony is cottony or wooly

Picture 13 - Saksinaea vasiformis culture

ASCOMYCETES
LABORATORY IDENTIFICATION

Aspergillus – 4 species commonly isolated:
1. A. fumigatus
2. A. flavus
3. A. niger
4. A. terreus

- Rapidly growing
- Septate hyphae
- Specialized hyphal segment known as a foot cell and serves as the base of origin of the conidiophores
- Conidiation occurs in a specialized fruiting body composed of swollen vesicle
- Identification is based on the microscopic and culture morphology

The 4 common species of Aspergillus:

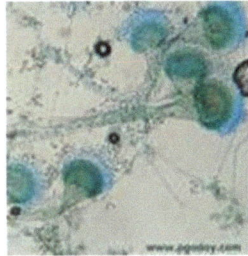

Picture 14 – A. fumigatus

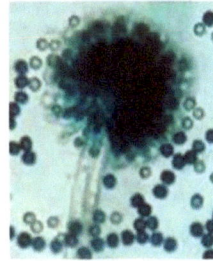

Picture 15 -A. niger

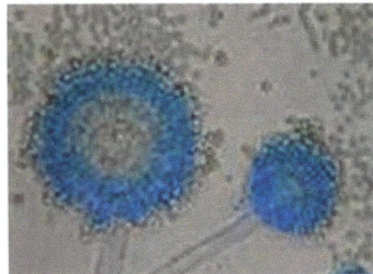

Picture 16 - A. flavus

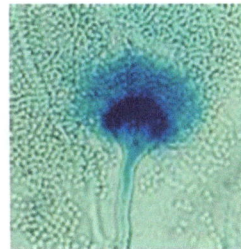

Picture 17 - A. Terreus

Picture 18 - Aspergillus colony morphologies

- Colony morphology: white to blue to green and tend to turn gray with age

Fusarium species
- Cottony on Potato Dextrose Agar (PDA)

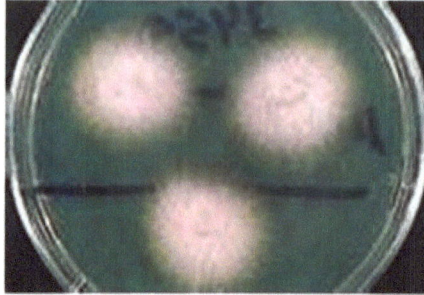

Picture 19 - Fusarium oxysporum

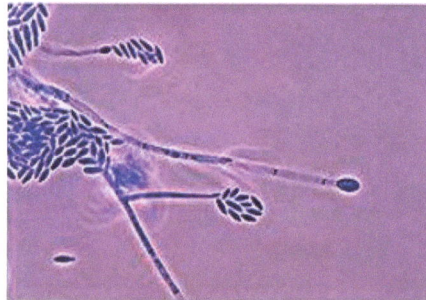

Picture 20 - Fusarium verticillioides

BASIDIOMYCETES
LABORATORY IDENTIFICATION

Alternaria species
- Dematiaceous (dark color) mold
- Colony is dark gray, brown, or black, wooly hair or velvety surface colonies
- Black pigmentation on the reverse of the agar plate

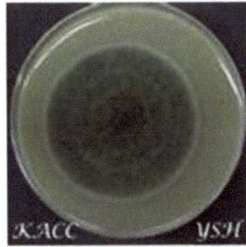

Picture 21 - Alternaria spp.

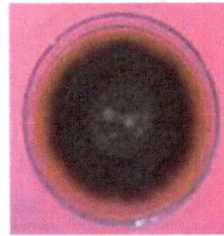

Picture 22 - Alternaria spp.

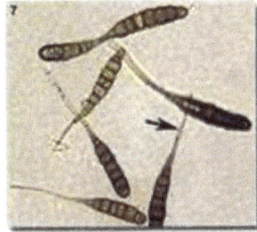

Picture 23 - Alternaria spp.
50X

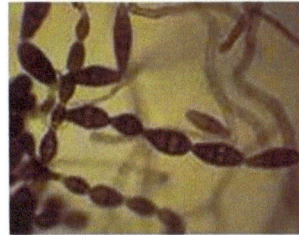

Picture 24 - Alternaria spp.
50X

Tinea – general term use to describe skin mycoses, usually caused by dermatophytes.

- Tinea versicolor or Ptyriasis versicolor– caused by Malassezia furfur, a yeast infection resulting in hypopigmentation of the skin ("ring-worm"-like). Identification is through direct skin scrapings, the structures resemble "sausage and meatballs" or "spaghetti and meatballs" under the microscope. Growth requires lipid or olive oil.

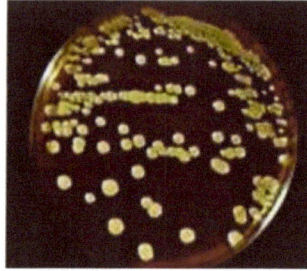

Picture 25 – Malassezia furfur culture

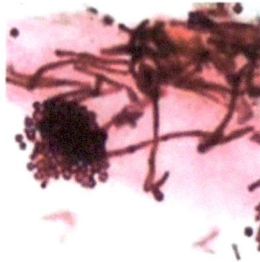

Picture 26 – Malassezia furfur 50X

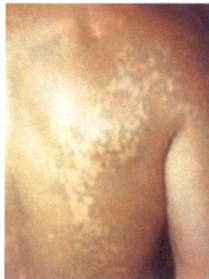

Picture 27 – Ptyriasis versicolor

- Tinea nigra palmaris – caused by Exophiala werneckii, a mold

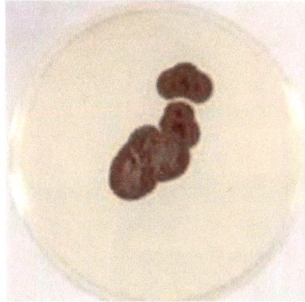

Picture 28 – Exophiala werneckii culture

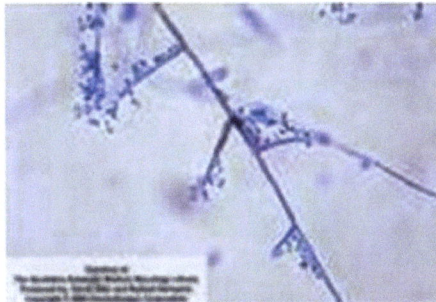

Picture 29 – Exophiala werneckii 50X

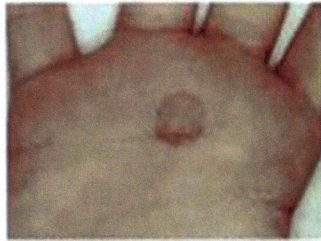

Picture 30 – Tinea nigra palmaris

Other Tinea:
- Tinea capitis – scalp, eyebrow, eyelashes
- Tinea pedis – foot
- Tinea unquium – nails
- Tinea corporis – ring worms of the body

- Tinea cruris – groin, perineum
- Tinea barbae - beard
- Piedra – infection of hair
- Black piedra – nodules on hair shaft composed of fungal elements caused by Peidraia hortae

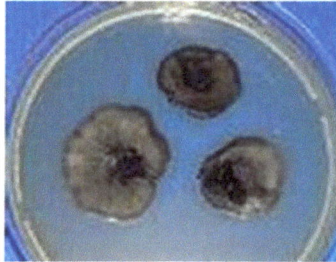

Picture 31 – Piedraia hortae/nigra culture

Picture 32 – Piedraia hortae microscopic

Picture 33 – Hair nodule infected with P. hortae

- White piedra – soft white to brown nodule of intertwined hyphae on hair shaft, fragments to yeast-like arthrospores caused by Trichosporon beigelii

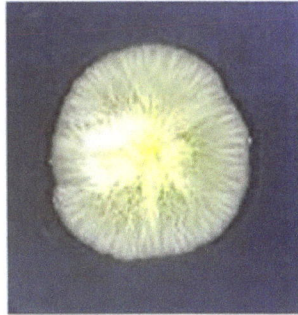

Picture 34 – Trichosporon spp. culture

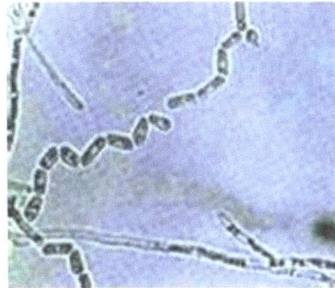

Picture 35 – Trichosporon beigelii 50X

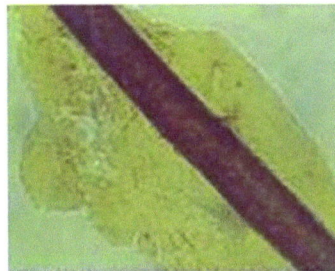

Picture 36 – White piedra of the hair

Continuation Ascomycetes - SUPERFICIAL MYCOSIS

- Dermatophytes – group of 3 types of fungus that commonly causes skin diseases in animals and humans.
- Dermatophyte Testing Medium (DTM) – sabureaud dextrose agar with cyclohexamide and chloramphenicol – will differentiate between Dermatophytes (produces red pigment) and non-dermatophytes (may be inhibited by cyclohexamide or if it grows does not produce red pigment)

1. Microsporum - causes tinea capitis and tinea corporis

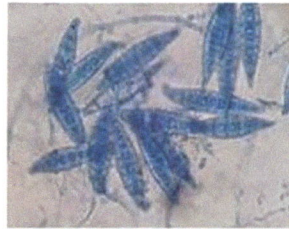

Picture 37 – Microsporum canis 50X

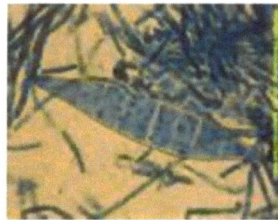

Picture 38 – Microsporum gypseum 50X

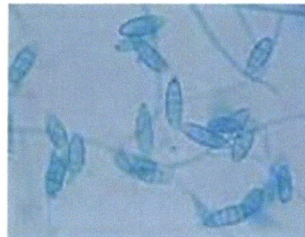

Picture 39 – Microscporum audouinii 50X

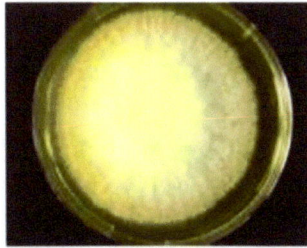

Picture 40 – *Microsporum canis* culture

Microsporum audouinii – chronic epidemic tinea capitis in children, white-tan colonies with yellow-red-brown reverse

Microsporum canis – acute epidemic tinea capitis in children

Microsporum gypseum – acute inflammatory tinea corporis, white-tan colonies with red-brown on reverse

- ▫ Colonies grow rapidly as white-tan with yellow green lemon color at the periphery, reverse side is yellow and changes as it grows older

- ▫ Microscopically forms both macroconidia and microconidia. Macroconidia are hyaline, multiseptate, fusiform.

2. Epidermophyton – causes tinea corporis (ringworm), tinea cruris (jock itch), tinea pedis (athlete's foot), onychomycosis or tinea unguium (fungal infection of the nail bed)

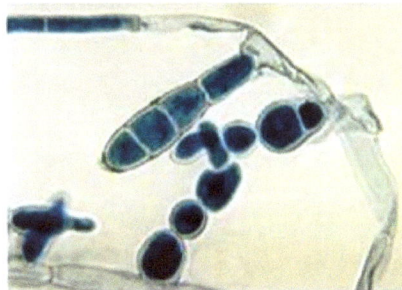

Picture 41 – *Epidermophyton flocossum* 50X

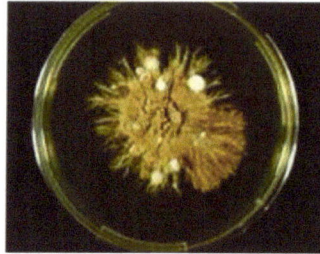

Picture 42 – Epidermophyton flocossum culture

- Colonies appear white-tan and olive-khakii with age
- Microscopic - smooth, club-shaped or finger-like thin-walled macroconidia which are often produced in clusters growing directly from the hyphae, no microconidia.

3. Trichophyton – most common cause of athlete's foot, jock itch and ringworm.

- Colony morphology of most Trichophytons are either granular or velvety, the only one with pigment (red) is Trichophyton rubrum (but still some may not produce a pigment)
- Microscopically, most Trichophytons have smooth-walled macroconidia (which is rare), a cigar-shaped and spherical microconidia
- The special media to identify the different dermatophyte species include Trichophyton Agar1 (no thiamine), Trichophyton Agar 4 (with thiamine) Urea Agar, Potato Dextrose Agar (PDA), Rice grain medium (M. Audouinii will not grow).

Picture 43 – Trichophyton mentagrophytes culture

Trichophyton mentagrophytes – acute tinea pedis, crème-tan colonies and tan-brown and dark red on reverse, microscopic shows spherical microconidida in singles or clusters on branched conidiospores

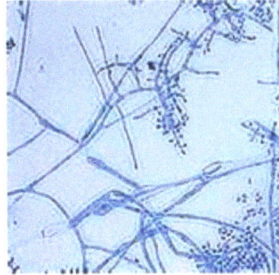

Picture 44 – T. mentagrophytes (50X)

Trichophyton rubrum – chronic tinea pedis, unguim and pedis, white velvety colonies with red-wine pigment. Microscopic – macroconidia is cigar-shaped, microconidia is tear-dropped shapes

Picture 45 – Trichophyton rubrum (50X)

Trichophyton tonsurans – tinea capitis in adults and children, colonies are highly variable, from white/creamy, yellow, tan and pink. Microscopic –microconidia vary in size and shape

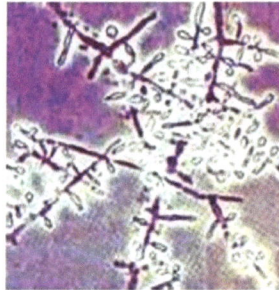

Picture 46 – Trichophyton mentagrophytes (50X)

Trichophyton violaceum – causes chronic non-inflammatory finely scaling lesions of skin, nails beard and scalp. Colonies are waxy, irregular shaped with areas that have a dark violet color and other that is white, RARE MACRO AND MICROCONIDIA

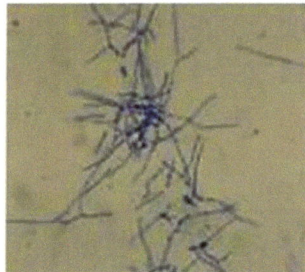

Picture 47 – Trichophyton violaceum (50X)

Trichophyton schoenleinii – favus (chronic skin infection), waxy appearance with tan-brown color and with age it becomes irregular with folden surfaces and hyphae is swollen giving a "nailhead" cultural morphology. Microscopic shows "favic chandeliers" or "antler-like" hyphae, RARE MACRO AND MICROCONIDIA.

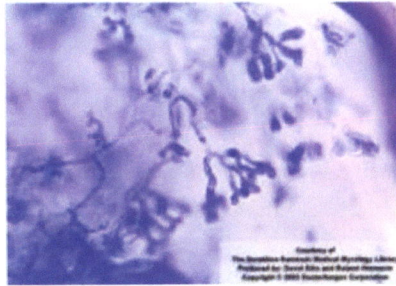

Picture 48 – Trichophyton schoenleinii (50X)

Trichophyton verrucosum – tinea barbae, colonies are white, creamy color. Microscopic shows many chlamydiospores that are sometimes referred to as "chain of pearls", microconidia is rare and small

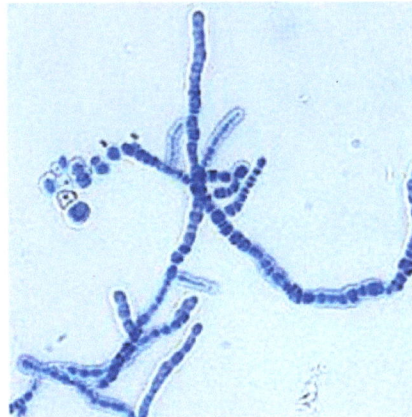

Picture 49 – Trichophyton verrucosum (50X)

SUBCUTANEOUS MYCOSES

1. Sporotrichosis – worldwide

Sporothrix schenkii – most frequently infects gardeners, basketmakers, florists, farmers and nurserymen. A dimorphic fungus that grows as mold at 25⁰C and yeast at 37⁰C.

- Macroscopic colony morphology as a yeast is moist, glabrous, white-to brown-black. Colony morphology as a mold is white and darkens to brown with age and has a folded tough appearance
- Microscopic morphology as a yeast is round-oval and pear-shaped conidia. Microscopic mosphology when it is a mold shows a "flowerette" arrangement

Picture 50 – Sporothrix schenckii culture

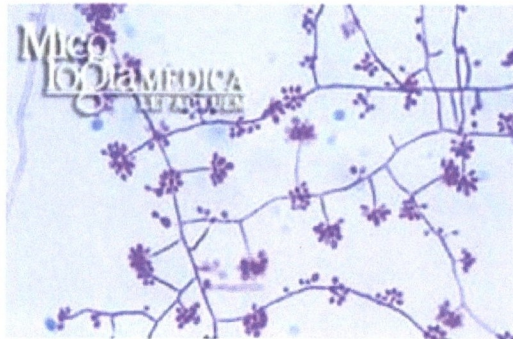

Picture 51 – Sporothrix schenckii microscopic mold form (50X)

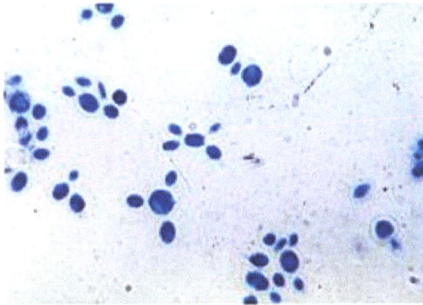

Picture 52 – Sporothrix schenckii microscopic yeast form (50X)

2. Maduramycosis (mycetoma) – subcutaneous infections in which the tissue is markedly swollen with the formation of deeply penetrating sinus tracts that break through the superficial skin and discharge purulent material. This infection is found worldwide and characterized with swelling, purplish discoloration, tumor-like deformities of subcutaneous tissue and multiple sinus tracts that drain pus containing granules.

Acremonium spp.

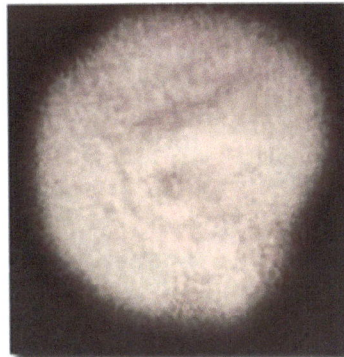

Picture 53 – Acremonium spp. culture

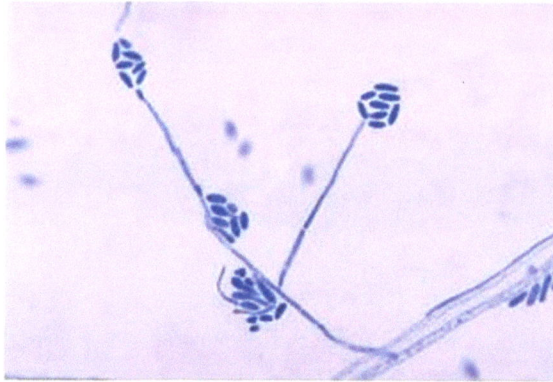

Picture 54 – Acremonium spp. (50X)

Exophiala jeanselmei

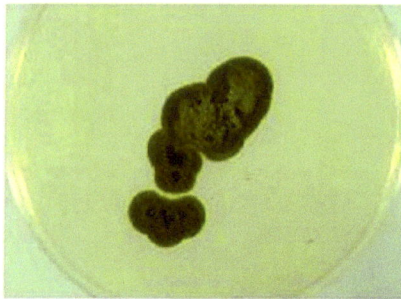

Picture 55 – Exophiala jeanselmei culture

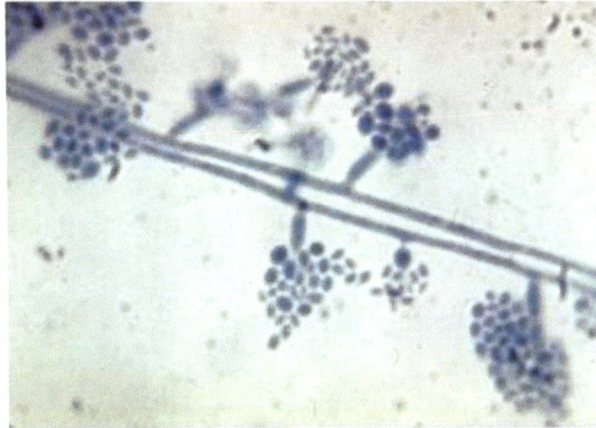

Picture 56 – Exophiala jeanselmei (50X)

Pseudoallescheria boydii – colonies grow rapidly, fluffy in appearance, white to moist gray in color and the reverse turn brown with age. The most common causative agent of Maduramycosis in the U.S. Microscopic – conidia may rise directly from the hyphae or from the tip of the conidiosphores.

3. Chromoblastomycosis – cutaneous and subcutaneous infection characterized by the formation of elevated, roughened verrucous vegetations and the infection may spread over the dorsal surfaces of the feet and lower leg. This infection is found worldwide but primarily seen in the tropics or subtropics.

    ◦ Most species that causes chromoblastomycosis are extremely slow growers and dematiaceous (dark) fungi
    ◦ Most species are filamentous at 25⁰C and at 37⁰C
    ◦ The genera and species are differentiated by the 3 types of sporulation

Cladosporium carrionii – microscopic morphologies include dark-pigmented and septate hyphae, elongated conidiospores that produce chains of ellipsoid and smooth-walled conidia

Picture 57 – Cladosporium carrionii culture

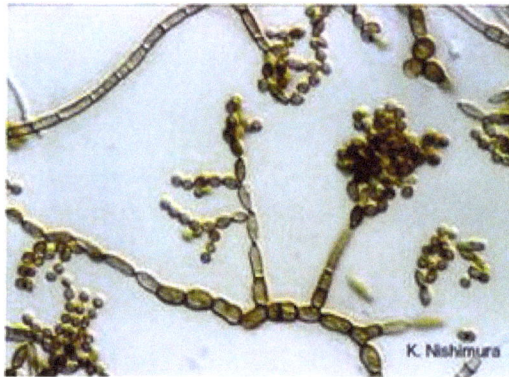

Picture 58 – Cladosporium spp. (50X)

Phialophora – microscopic morphology includes elongated conidiospores that are tubular and tapered with a narrow end with a smooth, ellipsoid conidia that can gather in clusters around the conidiospores

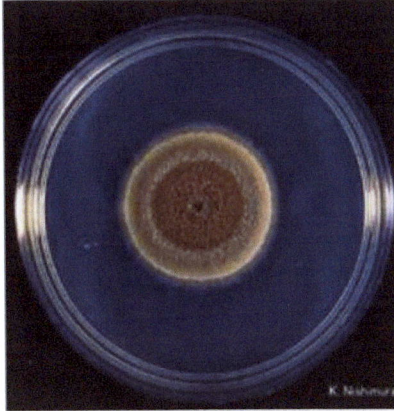

Picture 59 – Phialophora spp. culture

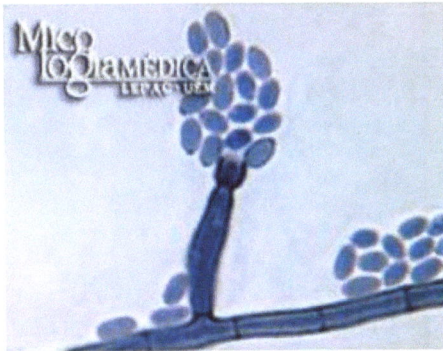

Picture 60 – Phialophora spp. (50X)

Fonsecaea pedrosoi – has a dark pigmented hyphae and conidia and the 3 types of conidial formation may be present (Phialophora, Cladosporium and Rhinocadiella)

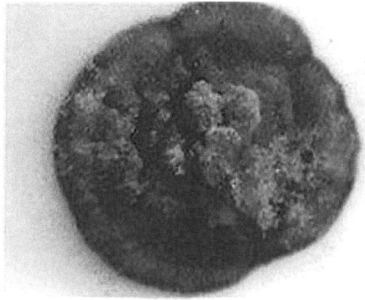

Picture 61 – Fonsecaea pedrosoi culture

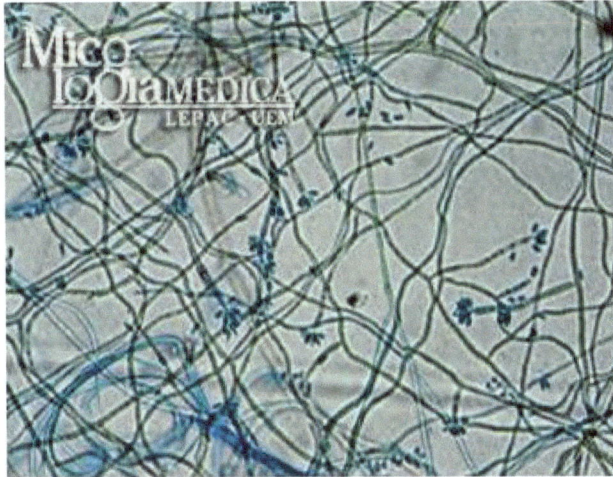

Picture 62 – Fonsecaea pedrosoi (10X)

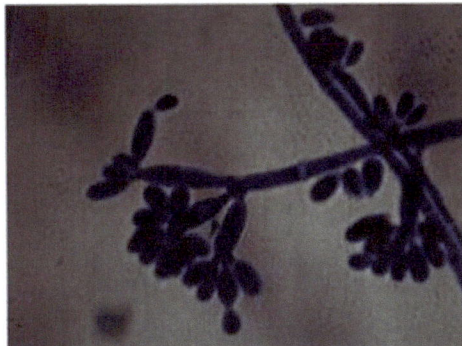

Picture 63 – Fonsecaea pedrosoi conidia (50X)

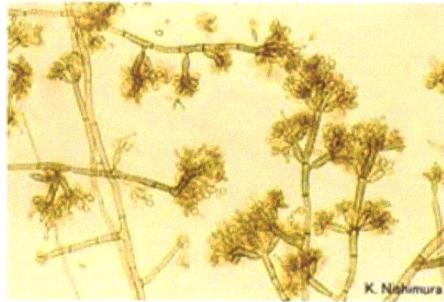

Picture 64 – Fonsecaea pedrosoi conidia (50X)

4. Phaeohyphomycosis – mycotic infection of humans and lower animals caused by a number of dematiaceous fungi where the tissue morphology differentiates it between other clinical types of diseases involving brown-pigmented fungi. When the tissue morphology of the organism is a grain the infection is mycotic mycetoma and when it is sclerotic body, the infection is chromoblastomycosis.

Alternaria spp.

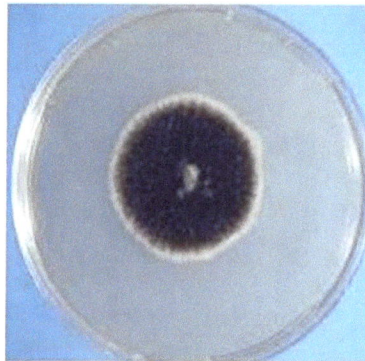

Picture 65 – Alternaria spp. culture

Picture 66 – Alternaria spp. (50X)

Cladosporium spp.

Exophiala spp.

Phialophora spp.

Wangiella dermatitidis

Picture 67 – Wangiella dermatitidis culture

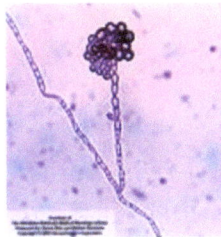

Picture 68 – Wangiella dermatitidis (50X)

SYTEMIC MYCOSES
- >90% of infections are symptomatic or for a very short duration
- Very virulent and caused by dimorphic fungi
- Infects otherwise healthy hosts and higher incidence in adult males
- Infection is through inhalation of spores
- Organisms are normal habitats of soil
- Infection is insidious (slow) and may not be diagnosed until autopsy
- Immunocompromised patients are of higher risk to contract the infection and have higher fatality

Coccidiodes immitis

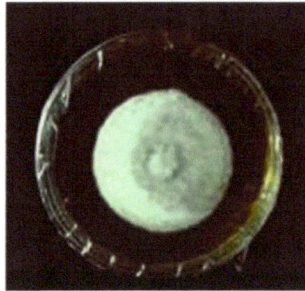

Picture 69 – Coccidiodes immits culture

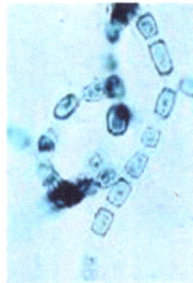

Picture 70 – Coccidiodes immitis mold form (50x)

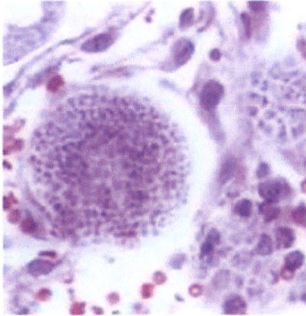

Picture 71 – Coccidiodes immitis yeast in tissue (100x)

Characteristics of Coccidioides immitis:

- Endemic in dry, arid soil of lower Sonoran, Western and South Western desert region of the U.S. and Kern County in California
- Causes pulmonary disease known as San Joaquin Valley Fever or Valley Fever
- Also causes cutaneous infection

LABORATORY IDENTIFICATION

- Dimorphic fungi, yeast at 37^0C and mold at 25-28^0C
- Colony morphology – Mold -white, fluffy and sporulates in 6-11 days
- Microscopic – Mold - septate hyphae and arthroconidia with alternating empty spaces. Yeast – only in tissue, mature spherules produces endospores by undergoing progressive cleavage
- ID by Exoantigen test
- ID by DNA probe
- ID by PCR

TREATMENT:

- Amphotericin B (95% susceptible)
- Voriconazole
- Caspofungin
- Fluconazole

Histoplasma capsulatum

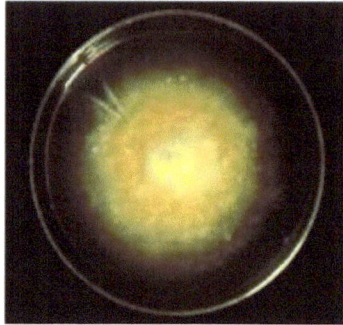

Picture 72 – Histoplasma capsulatum culture

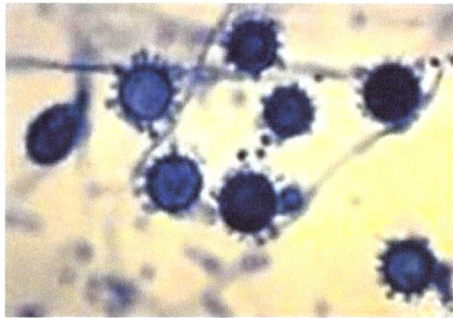

Picture 73 – Histoplasma capsulatum mold (50X)

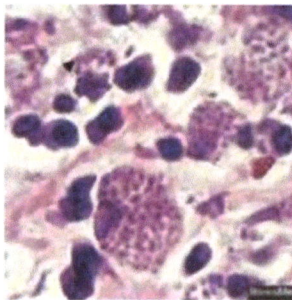

Picture 74 – Histoplasma capsulatum yeast (100x)

Characteristics of Histoplasma capsulatum:

- An obligate intracellular organism, reside in macrophages of the reticuloendothelial system
- Dimorphic, yeast at 37⁰C and mold at 25-28⁰C
- Found around the Mississippi-Ohio River Valley
- Source is bird droppings
- Causes pulmonary infection, severe fever, chest pain, chills and cough

LABORATORY IDENTIFICATION:

- Colony morphology: Mold – white-tan, slow-growing. Yeast – better to stain the tissue than to grow it
- Microscopic: Mold – tuberculate macroconidia and spherical or pyriform, microconidia on short lateral conidiophores. Yeast (in tissue or bone marrow) – round to oval, small, narrow-based, budding yeast cells stained with GMS (in tissue) and Calcofluor white stain (in bone marrow)
- ID by Exoantigen confirmation
- ID by DNA probe
- ID by PCR

TREATMENT:

- Amphotericin B
- Itraconazole
- Fluconazole
- Voriconazole
- Caspofungin

Blastomyces dermatitidis

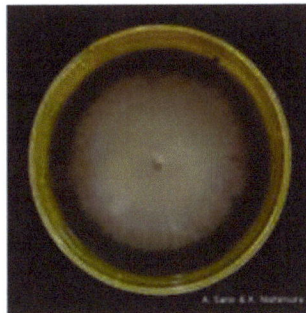

Picture 75 – Blastomyces dermatitidis culture

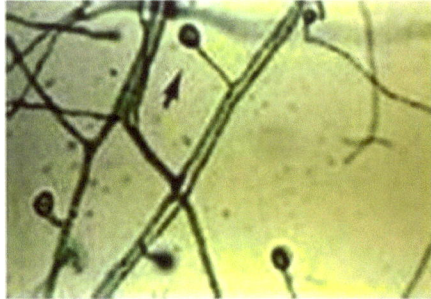

Picture 76 - Blastomyces dermatitidis mold (50X)

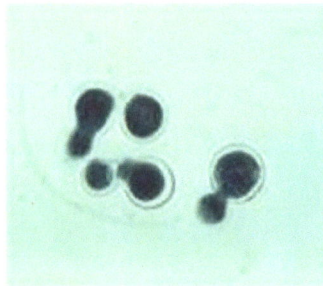

Picture 77 - Blastomyces dermatitidis yeast (100X)

Characteristics of Blastomyces dermatitidis:
- Causes pulmonary, skin, bone, urogenital infections
- Dimorphic: Yeast at 37^0C and mold at $25\text{-}28^0$C

LABORATORY IDENTIFICATION:
- Colony morphology:
  Mold –variable in morphology and rate of growth, may be rapid-growing, fluffy, white or slow-growing, glabrous tan colonies.
  Yeast (in tissue) – better to stain tissue and look for the yeast.
- Microscopic: Mold – spherical, pyriform (pear-shaped) conidia on short, lateral conidiophores (lollipop-shaped microconidia). Yeast (in tissue) – large yeast with broad budding base
- ID by Exoantigen
- ID by DNA probe
- ID by PCR

TREATMENT:
- Amphotericin B
- Itraconazole
- Fluconazole
- Voriconazole
- Caspofungin

Paracoccidioides brasiliensis

Picture 78 – Paracoccidiodes brasiliensis culture

Picture 79 – Paracoccidiodes brasilienses mold (50X)

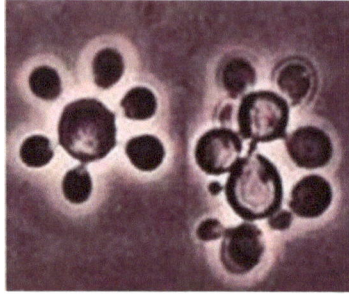

Picture 80 – Paracoccidiodes brasiliensis yeast (100X)

Characteristics of Paracoccidiodes brasiliensis:
- Found in Central and South American forests
- Causes pulmonary infection and ulcers of mucus membranes
- Dimorphic, yeast at 37^0C and mold at 25-28^{0}C

LABORATORY IDENTIFICATION:
- Colony morphology:
  Mold – white-tan, slow growing, waxy and velvety.
  Yeast – better to stain the tissue and look for the multiple budding yeast.
- Microscopic: mold – sterile hyphae (chlamydospores may be present). Yeast (in tissue) – large yeast with "mickey mouse" type of budding, a mother cell with multiple budding yeasts described as "spoke of wheels"

TREATMENT:
- Sulfonamides
- Amphotericin B
- Ketoconazole
- Itraconazole
- Fluconazole
- Terbinafine

Stains

- KOH-Calcofluor Exam – examine the specimen on a wet mount and look for fungal elements, hyphae, budding. For hair shafts, look for nodules and hyphae
- Gram Stain – examine for yeast which will stain large gram positive with or without budding
- India Ink – confirm Cryptococcus neoformans
- Gomori Methenamine Silver (GMS) – to identify a yeast-like fungus Pneumocystis jiroveci in the tissue

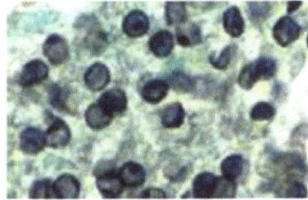

Picture 81 - GMS positive Pneumocystis jiroveci

- Lactophenol Cotton Blue – fungal elements absorbs the stain and could be observed on a wet mount

SECTION V: Medically Important Fungi

OPPORTUNISTIC FUNGI

- Pneumocystis – opportunistic pathogen seen in individuals with compromised immune system, anamorphic fungus
- Stachybotrys – normally grows on cellulose, found on dead plants, toxic spores can cause fatal pulmonary hemorrhage in infants
- Zygomycosis – disease most often occurs in immunocompromised patients
  - Mucormycosis – caused by Rhizopus and Mucor, infection occurs mostly in patients with Diabetes Mellitus, leukemia or undergoing treatment with immunosuppressive drugs
- Aspergillosis – caused by Aspergillus, occurs in people with debilitating lung diseases or cancer and have inhaled Aspergillus spores
- Cryptococcus and Penicillum – can cause fatal diseases in AIDS patients
- Candida albicans – causes yeast infection or candidiasis, may occur as vulvovaginal candidiasis or thrush, a mucocutaneous candidiasis

ECONOMIC EFFECTS OF FUNGI

- Aspergillus niger – used to produce citric acid for foods and beverages
- Saccharomyces cerevisiae – used to make bread and wine
- Some fungus could be genetically engineered to produce a variety of proteins including hepatitis B vaccine
- Saccharomyces and Torulopsis – used as protein supplements for humans and cattle
- Trichoderma – used commercially to produce the enzyme cellulase used to remove plant cell walls to produce a clear fruit juice
- Taxomyces – produce taxol, an anti-cancer drug
- Entomorphaga – kills gypsy moths that destroys trees
- Candida oleophila – used to prevent fungal growth on harvested fruits
- Cryphonectria parasitica – kills chestnut trees
- Ceratocystis ulmi – Dutch elm disease

SECTION VI: Safety

Class II Biological Safety Cabinet must be used at all times when performing any mold testing or processing

Incinerator-burner – use to burn animal carcasses

Handling of discards
- Proper discard of all biological materials (specimens) and cultures
- Mold positives – tape up all the plates or treat with bleach then discard in red biohazard bags
- Yeasts – discard in biohazard red bag
- Bioterrorism precaution – Cocci spores

SECTION VII: Antifungal Susceptibility Testing (AST)

ANTIFUNGAL DRUGS

- Agents affecting Fungal Sterols – target sterols in fungal plasma membrane (ergosterols), when biosynthesis of ergosterols is interrupted, membrane becomes excessively permeable killing the cell

    a. Polyenes
    - Amphotericin B – use for systemic fungal diseases (histoplasmosis, coccidiomycosis and blastomycosis), causes drug toxicity particularly to the kidneys
    b. Azoles
    - Imidazoles – 1st azoles (Clotrimazole and miconazole) – for topical application of cutaneous mycoses (athlete's foot and vaginal yeast infections)
    - Ketoconazoles – orally, less toxic alternative to amphotericin B for systemic fungal infections, ointments are used to treat dermatomycoses
    - Voriconazole – promising new broad spectrum antifungal and expect to replace amphotericin B for treatment of systemic antifungal infections, use in treatment of aspergillosis of the CNS because of its ability to penetrate blood-brain barrier
    - Triazole antibiotics (fluconazole and itraconazole) – less toxic, water soluble
    c. Allylamines (terbinafine and naftifine)– inhibit the biosynthesis of ergosterols in a manner that is functionally distinct, frequently used when resistance to azole-type antifungals arises

- Agents Affecting Fungal Cell Walls – primary target is β -glucan, inhibition results in lysis of the fungal cell

    a. Echinocandins – inhibit biosynthesis of glucans
    - Caspofungin (Candidas) a member of the echinocandin group, it combats Aspergillus infections in persons whose immune system is compromised, also effective against Candida spp. and Pneumocystis jiroveci

- Agents Inhibiting Nucleic Acids

    a. Flucytocine – analog of pyrimidine cytocine, fungal cells will be able to convert flucytocine to 5-fluorouracil which gets incorporated in the RNA and disrupts proteins synthesis. Since humans cannot convert this drug, it does not cause harm to humans but it may be toxic to the kidneys and bone marrow

- Other Antifungal drugs –
    a. Griseofulvin – produced by Penicillum, active against superficial dermatophytic fungal infections of the hair and nails, taken orally, drug binds selectively to the keratin found in the skin, hair follicles and nails, blocks microtubule assembly which interferes with mitosis and inhibits fungal reproduction

  b. Tolnaftate – alternative to miconazole as a topical agent for the treatment of athlete's foot

    c. Undecylenic acid – fatty acid with antifungal activity against athlete's foot, not as effective as tolnaftate and imidazoles

  d. Pentamidine isethionate – used in treatment of Pneumocystis pneumonia, appears to bind DNA

MYCOLOGY-WRITTEN EXAMINATION

Student Name_____

(Please circle the correct answer)

1. Cryptococcus neoformans is an important pathogen among AIDS and immunocompromised patients and could be identified by the following tests except:

   a. India ink

   b. Urea

   c. Cornmeal agar

   d. Niger seed agar

2. A dimorphic mold that frequently infects gardeners, basket makers, florists, farmer and nursery workers

   a. Trychosporon beigelii

   b. Acremonium spp.

   c. Exophiala jeanselmei

   d. Sporothrix schenkii

3. Type of fungal infection that infects only the hair, skin and nails and mainly caused by dermatophytes:

   a. Superficial mycoses

   b. Cutaneous mycoses

   c. Subcutaneous mycoses

   d. Systemic mycoses

4. Name the test in the picture _____

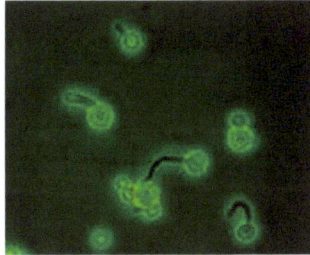

5. This antifungal is produced by Penicillum and is active against superficial dermatophytic fungal infections

   a. Undecylenic acid

   b. Voriconazole

   c. Griseofulvin

   d. Amphotericin B

6. A mold that causes this type of skin infection:

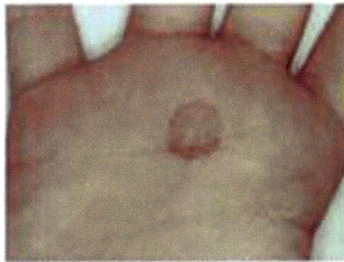

Figura 1: Mancha acastanhada, com 1.5cm de diâmetro, bianco perciso, na palma da mão direta / *Figure 1. Brown stain, 1.5 cm in diameter, well-defined borders on the palm of the right hand*

   a. Exophiala werneckii

   b. Piedra hortae

   c. Trichosporon beigelii

   d. Sporothrix schenkii

7. The antibiotic in the BHI with C and C are:

   a. Cyclohexamide

   b. Clarythromycin

   c. Clindamycin

   d. Chloramphenicol

   e. a and b

   f. a and c

   g. a and d

   h. b and c

   i. b and d

   j. c and d

8. The following are common pathogens isolated in the hair, skin and nails except:

   a. Microsporum canis

   b. Trichophyton rubrum

   c. Blastocytis dermatitidis

   d. Epidermophyton flocossum

9. The following molds causes maduramycosis except:

   a.   Acremonium spp.

   b.   Aspergillus spp.

   c.   Exophiala jeanselmei

   d.   Pseudoallescheria boydii

10. This antifungal inhibits nucleic acid synthesis

   a.   Griseofulvin

   b.   Pentamidine isethionate

   c.   Flucytosine

   d.   Caspofungin

11. Tinea versicolor is an infection resulting in hypopigmentation of the skin and is caused by this organism:

   a.   Cryptococcus neoformans

   b.   Malassezia furfur

   c.   Tinea nigra Palmaris

   d.   Alternaria spp.

12. The following are dimorphics except:

   a.   Histoplasma capsulatum

   b.   Coccidiodis immitis

   c.   Paracoccidioides brasiliensis

   d.   Cladosporium carionii

   e.   Blastomyces dermatitidis

13. The following are common pathogens isolated in the lungs except:

    a. Coccidiodes immitis

    b. Aspergillus fumigatus

    c. Trychophyton tonsurans

    d. Histoplasma capsulatum

14. This mold is used to make bread and wine

    a. Saccharomyces cerevisiae

    b. Trichoderma spp.

    c. Ceratocystis ulmi

    d. Aspergillus niger

15. The following molds causes phaeohyphomycosis:

    a. Wangiella dermatitidis

    b. Alternaria spp.

    c. Fonsecaea pedrosoi

    d. Phialophora spp.

16. The following fungus are capable of reproducing by budding except:

    a. Torulopsis glabrata

    b. Candida albicans

    c. Aspergillus fumigatus

    d. Cryptococcus neoformans

17. This stain will identify a yeast-like fungus Pneumocystis jiroveci in the tissue:

    a. India Ink

b. Gomori Methanamide Silver (GMS)

c. Lactophenol Cotton Blue

d. Calcofluor Stain

18. This mold normally grows on cellulose and found on dead plants and is known as "toxic mold" causing fatal pulmonary disease:

a. Rhizopus spp.

b. Aspergillus spp.

c. Coccidiodes immitis

d. Stachybotris spp.

19. Fungal infection deep within the body caused by dimorphic fungus:

a. Subcutaneous mycoses

b. Cutaneous mycoses

c. Mucocutaneous mycoses

d. Systemic mycoses

20. These molds causes mucormycoses that occurs in patients with Diabetes Mellitus, leukemia or undergoing treatment with immunosuppressive drugs

a. Absidia

b. Rhizopus

c. Mucor

d. Saksenaea

e. a and b

f. a and c

g. a and d

h. b and c

i. b and d

j. c and d

21. This mold kills gypsy moths that were destroying trees:

    a. Taxomyces

    b. Entomorphaga

    c. Trichoderma

    d. Candida oleophila

22. This fatal systemic infection is caused by a dimorphic fungi with characteristic septate hyphae and arthroconidia with alternating empty spaces

    a. Histoplasma capsulatum

    b. Coccidioides immitis

    c. Blastocystis hominis

    d. Paracoccidioides brasiliensis

23. The following media will yield differing yeast morphology:

    a. Cornmeal Agar

    b. Potato Dextrose Agar

    c. Sabouraud's Dextrose Agar

    d. Brain Heart Infusion Agar

24. Fungus grows better at a pH of:

a. Acidic

b. Basic

c. Neutral

d. Severely basic

25. The following molds causes chromoblastomycosis except:

a. Fonsecaea

b. Cladosporium carionii

c. Exophiala jeanselmei

d. Phialophora spp.

26. The species of Candida that has green colonies in the cornmeal agar is:

a. Candida glabrata

b. Candida albicans

c. Candida tropicalis

d. Candida krusei

27. This antifungal agent is used for systemic fungal disease

a. Imidazole

b. Amphotericin B

c. Terbinafine

d. Tofnaftate

28. Fungus have cholesterol in their plasma membrane and humans have ergosterols

a. True

b. False

29. Fungal phylum composed of aseptate molds:

    a. Ascomycota

    b. Zygomycota

    c. Basidiomycota

    d. Glomeromycota

30. The following azoles are used for topical applications of athlete's foot and vaginal yeast infection

    a. Ketoconazole

    b. Clotrimazole

    c. Voriconazole

    d. Miconazole

    e. a and b

    f. a and c

    g. a and d

    h. b and c

    i. b and d

    j. c and d

Student Signature_____

Date_____

Total Correct Score_____

%_____

Answer Key Mycology Written Examination
30 Points

1. c
2. d
3. b
4. Germ tube
5. c
6. a
7. g
8. c
9. b
10. c
11. b
12. d
13. c
14. a
15. c
16. d
17. b
18. d
19. d
20. h
21. b
22. b
23. a
24. a
25. c
26. b
27. b
28. b
29. b
30. i

Afterword

In a perfect world, after their clinical rotation, CLSs and MLTs will be working in all four areas of study of medical technology (Microbiology, Chemistry, Hematology and Immunohematology), retain all that they know and live and work happily ever after. But as we all know that is far from the case, more frequently than not, a CLS or MLT will be stuck in one or two specialized area of study. Mycology is a unique sub-subspecialty of Microbiology that both the macroscopic and microscopic part of it require years of experience for a CLS to feel a high level of confidence. CLSs who had been away from the Mycology department for a period of time or never had a chance to work in the Mycology department will need a good refresher course before venturing into this department. This manual does not claim to be able to boost someone's confidence overnight or claim to have all the answers, but instead this manual serves as a guide to re-discovering what one previously knew.

My hope is that this manual will serve its purpose and be a source of confidence to those who are brave enough to venture out in the mycology department as a newbee or someone who had been away from it for a period of time.

Glossary

This manual is directed towards CLS and MLT students at or nearing their clinical rotation as well as CLSs that had been away from the area of Mycology for a period of time and should already be familiar with most of the terms used in this manual. Only a selected number of terms below were chosen for further explanation.

ARTHROCONIDIA – also called arthrospore, refers to one of the small conidia borne in chains by various fungi.

ANAMORPHIC – refers to the inability of an organism to progress or change form from one type to another.

CHITIN – refers to a horny polysaccharide that forms part of the harder outer integument especially of insects, arachnids and crustaceans

CONIDIOSPORE – also called conidia, refers to the asexual spore produced on a conidiophore.

CUTANEOUS – refers to the skin

DORSAL – refers to a location near, on or toward the upper surface of an animal opposite the lower or ventral surface.

DORMANT – inactive stage of an organism.

EUKARYOTE – refers to the group of organisms that have cells containing visibly evident nuclei and organelles.

EXOANTIGEN – refers to a rapid and specific immunological test to identify an organism's antigen using a known antibody that is commercially prepared.

FUSIFORM – refers to a specific shape that tapers at the end.

GLABROUS – means smooth, a type of term to describe a colony or growth.

GLUCANS – refers to any polysaccharide composed only of recurring units of glucose.

MYCETOMA – refers to a condition marked by invasion of deep subcutaneous tissues with fungi or actinomycetes.

NOSOCOMIAL – refers to an infection that is hospital-acquired.

PYRIFORM – also means pear-shaped.

SKIN MYCOSES – refers to skin infection with a fungus.

SUBCUTANEOUS – refers to the area located under the skin.

STEROLS – refers to any of the various steroid alcohols widely distributed in animal and plant lipids.

TINEA CAPITIS – refers to the fungal infection of the scalp characterized by bald patches.

TINEA PEDIS – refers to the fungal infection of the feet.

TINEA BARBAE – refers to the fungal infection of the face and neck.

TUBERCULARE – means covered with nodules.

UNGUIUM – refers to the fungal infection of the nails, especially toenails.

V FACTOR – refers to growth factors Nicotinamide Adenine Dinucleotide (NAD) or Nicotinamide Adenine Dinucleotide Phosphate-Oxidase (NADPH) that is usually added to a medium to grow organisms with complex nutritional requirements.

X FACTOR – refers to the growth factor hemin that is usually added to a medium to grow organisms with complex nutritional requirements.

References

Brown, A. E. (2009). *Benson's Microbiological Applications: Laboratory Manual in General Microbiology, Eleventh Edition,* Boston, Massachusetts: WCB McGraw-Hill Companies.

Isenberg, H., Ed. (2004). *Clinical Microbiology Procedures Handbook, 2nd Edition*, New Hyde Park, New York: ASM Press.

Martinez, RL, Tovar, LJM, Gayoso, PM and Hernandez, FH. (2009). *Principios de Micologia Medica.* 1st ed. Mexico, D.F.

http://www.microbelibrary.org/about/index.asp?bid=1088. Accessed images May 14, 2011.

Merriam-Webster, m-w.com

About the Author

Mary Michelle Shodja earned her Bachelor of Science Degree in Medical Technology in 1992 from California State University (CSU) Dominguez Hills in Carson, California. She took her medical technology clinical year training from the Southern California Kaiser Permanente Medical Hospital and Regional Laboratory. She earned her Masters of Science Degree in Bioanalysis in 1995 from her alma mater CSU Dominguez Hills.

She gained Certifications in both the MLS American Society of Clinical Pathologists (ASCP) and CLS National Credentialing Agency (NCA) in 1993. Immediately after passing her California License, she started as a Staff CLS in the Microbiology Department of the Southern California Kaiser Permanente Regional Laboratory. For over 17 years as a CLS, she maintained a full-time position in the area of Microbiology and a part-time position as a generalist working in Hematology, Chemistry, Serology, Immunology and Non-transfusion Blood Bank.

Over the years, she took on managerial, supervisory, teaching and other administrative and consultative work but her real passion lies in the clinical bench work and teaching. She believes that in order to find what you're looking for, you have to try out everything else.

www.ingramcontent.com/pod-product-compliance
Lightning Source LLC
Chambersburg PA
CBHW041450210326
41599CB00004B/196